The Football Math
The Christmas Match

Adrian Lobley

To Sebastian

With thanks to:
Sebastian Wraith-Lobley
Anne Lobley
Sarah Wraith
Kaden Nicholson
Callan Nicholson
Asher Nicholson
Mrs Hayton
Finlay Cronshaw
Joshua Clements
Christopher Chamberlain
Cameron Stewart
Steph Stewart

Front/back cover illustrations by:
Alyssa Josue

 WORLD RANKINGS

At one point in the past, these were the best 10 countries at football.

Which team is ranked number 2 in the world?

Which position is France ranked?

	FIFA	
1		Brazil
2		Spain
3		Portugal
4		Netherlands
5		Italy
6		Germany
7		Argentina
8		England
9		France
10		Croatia

In the list, Brazil are said to be 'above' Spain, even though the number next to their name is lower

Which team is ranked 1 place above Germany?

Which team is ranked 2 places above Croatia?

Which team is ranked 2 places below Spain?

Which team is ranked 6 places below Brazil?

1

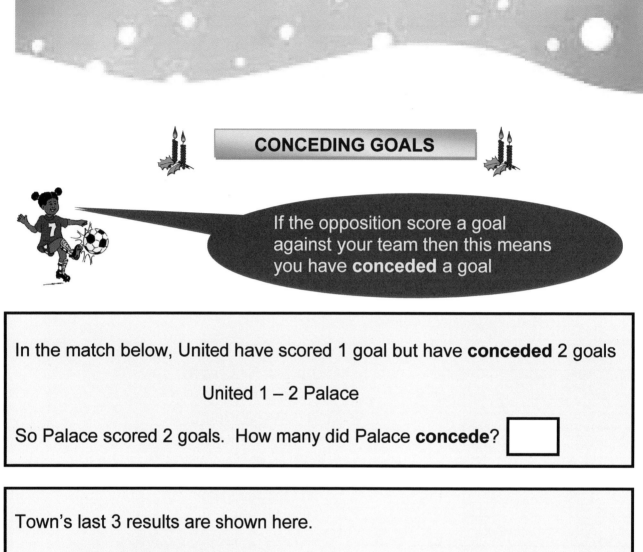

CONCEDING GOALS

If the opposition score a goal against your team then this means you have **conceded** a goal

In the match below, United have scored 1 goal but have **conceded** 2 goals

United 1 – 2 Palace

So Palace scored 2 goals. How many did Palace **concede**? ☐

Town's last 3 results are shown here.

Town 4-3 Argyle

Town 3-2 Palace

Town 0-0 Wanderers

How many goals have Town scored in total? ☐

How many goals have Town **conceded** in total? ☐

LEAGUE TABLES

A team gets:
3 points if they have won,
1 point if they have drawn,
0 points if they have lost

In the table below, work out how many points each team have got.

Enter your answers in the white boxes at the end of each line.

Team	Played	Won	Drawn	Lost	Points
Town	2	2	0	0	
City	2	1	0	1	
Rovers	2	1	0	1	
Albion	2	0	0	2	

Hint: Town have won 2 matches, so they receive 3 points for each match. This means Town have 3+3 points. Enter the answer to this sum in their white box.

Multiples of 4 include: 4, 8, 12, 16, 20, 24

Start at the bottom of the grid and draw a route to goal using **multiples of 4**.

4	13	19	8	15	30
20	9	15	33	19	31
24	12	8	20	12	16
17	13	7	9	5	4
99	26	2	8	16	24
22	2	3	12	9	17

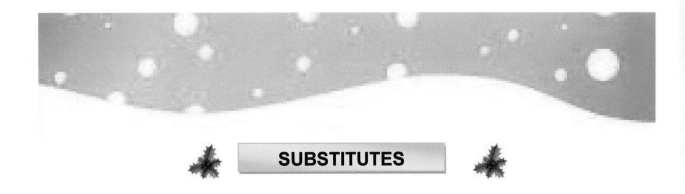

Below is a Squad of 8 players

In a 5-a-side match, only 5 of the Squad can be on the pitch at once

So 3 players will be sat out. These 3 are the Substitutes (Subs).

If a team have a Squad of 7 players and only 5 are allowed on the pitch, how many substitutes do they have?

If a team have a Squad of 9 players and only 5 are allowed on the pitch, how many substitutes do they have?

If a team have a Squad of 5 players and only 5 are allowed on the pitch, how many substitutes do they have?

GOAL DIFFERENCE

The difference between the number of goals scored and number of goals conceded, is called **Goal Difference**

Below, City won 3-1. They scored 2 goals *more* than the opposition, so City's **goal difference** for this match is **'Plus 2'**, written as **'+2'**.

Result	City's Goal Difference
City 3-1 Albion	+2

Below, City lost 1-3. They scored 2 goals *less* than the opposition, so City's **goal difference** for this match is **'Minus 2'**, written as **'-2'**.

Result	City's Goal Difference
City 1-3 Town	-2

Enter the goal difference for each of <u>City's</u> next matches below:

Result	Goal Difference
City 2-1 Town	☐
City 1-2 United	☐

Remember: You need to put a '+' or '-' in front of the number

GOAL DIFFERENCE

City's Goal Difference in each match has been calculated – but 2 are wrong!

Put 'X' next to each box that is wrong

Result			**City's Goal Difference**
City	3-1	Vale	+2
City	1-3	Palace	-2
City	4-1	Wanderers	+3
City	2-3	Ham	-1
City	2-1	North End	+5
City	3-5	Albion	-2
City	1-4	United	+6
City	1-1	Town	0

7

LEAGUE PLACINGS

In the table below, Town have the same number of **points** as City.

How many points do they both have?

Team	Played	Won	Drawn	Lost	Points	Goal Difference
Rovers	2	2	0	0	6	+4
City	2	1	0	1	3	+1
Town	2	1	0	1	3	-2
Albion	2	0	0	2	0	-4

The more points a team has, the higher they are placed. If teams have the same number of points, like City and Town, then the team with the best **Goal Difference** is placed the highest.

In the table, what is City's **Goal Difference**?

In the table, what is Town's **Goal Difference**?

Who has the best Goal Difference: City or Town?

So that's why City are above Town in the table

THE FOOTBALL MANAGER

You are a football manager who wants to build a 5-a-side team. You have no players at the moment, so you need to buy 5 players.

There are 10 players to choose from below. Circle 5 you want in your team.

Cole
3

Claus
5

Singh
4

Dent
4

Bond
3

White
2

Hill
1

Kidd
5

Ali
2

Snow
4

The numbers under each player are the ratings of how good they are.

> 'Kidd' is brilliant - he has a rating of 5
> 'Hill' is not as good - his rating is 1

To find out how many goals your team scored in your first match, add together the ratings of your 5 players ☐

To find out how many goals the opposition scored, add together the ratings of the other 5 players. ☐

Did your team win?

Multiples of 4 include: 4, 8, 12, 16, 20, 24

Start at the bottom of the grid and draw a route to goal using **multiples of 4**

23	13	19	9	15	8
33	9	15	33	19	20
25	12	20	24	14	24
14	4	7	12	20	12
7	8	2	11	17	31
22	4	3	40	9	17

Below, half a football plus another half a football, equals 1 whole football. Can you add the others?

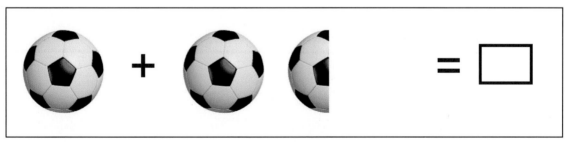

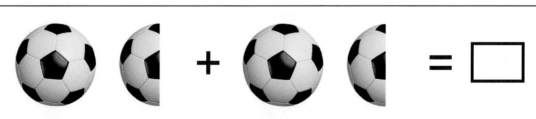

11

POINTS TALLY

Your team are playing 5 matches. You have won the first match 3-1.

To find out your next 4 scores you will need a die.

Roll the die 4 times and write the 4 numbers in the red squares below.

3 - 1		3 points
☐ - 4		__ points
☐ - 2		__ points
☐ - 0		__ points
☐ - 1		__ points

As your team won their first match 3-1, then you received 3 points for that match

For each of the other matches you have won, write '3' points next to it.

For each match you have drawn, write '1' point next to it.

For each match you have lost, write '0' points next to it.

How many points did you get in total? ☐

12

Penalty!

The opposition team have been given a penalty!   Use the clues below to work out which square the opposition striker kicks the ball towards.

Cross out the numbers in the picture, after answering each clue:

- He **doesn't** shoot at any number that is a multiple of 5
- He **doesn't** shoot at any number that is greater than 8
- He **doesn't** shoot at any even numbers
- He **doesn't** shoot at any number less than 4

Enter the number of the square where the opposition striker shoots

Football Shirts

Here we are going to "**Round up**" numbers to the next multiple of 10.

Multiples of 10 include:
10, 20, 30, 40, 50, 60, 70, 80…

So for each shirt number below, keep counting upwards, but stop when you get to a multiple of 10. The first 2 have been done.

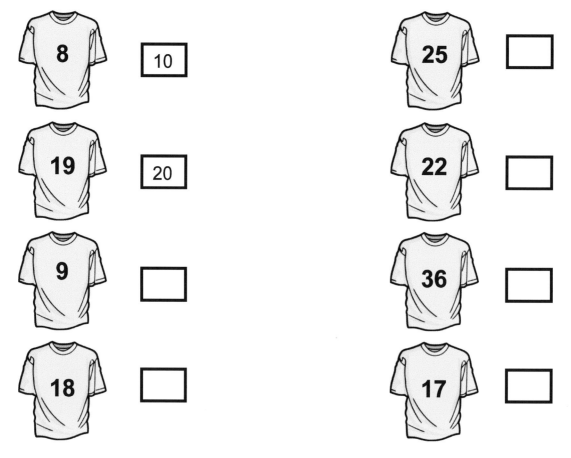

Football Shirts

Multiples of 10 include:
10, 20, 30 ,40, 50, 60, 70, 80…

This time "**<u>Round down</u>**" the numbers on the kits to the nearest multiple of 10.

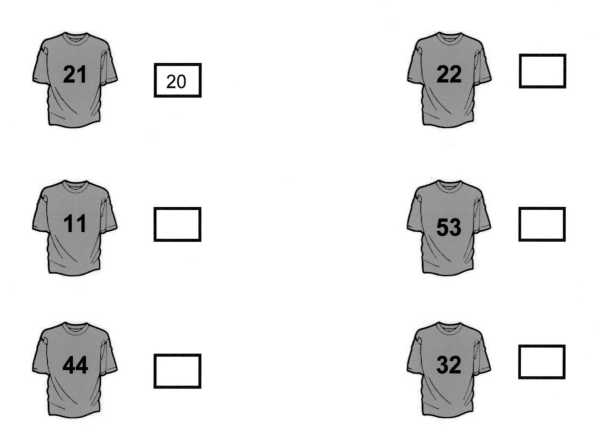

21 → 20

22 → ☐

11 → ☐

53 → ☐

44 → ☐

32 → ☐

 PITCH MEASUREMENTS

In football you will hear people say the term, '6 yard box'.

Can you spot where the '6 yard box' is on the football pitch below?

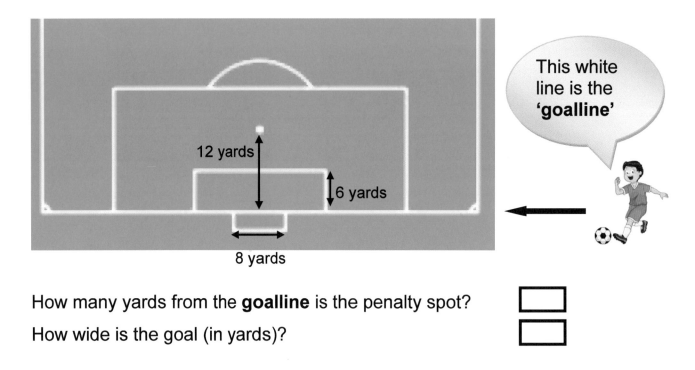

This white line is the **'goalline'**

12 yards

6 yards

8 yards

How many yards from the **goalline** is the penalty spot?

How wide is the goal (in yards)?

A 'yard' is an old English measurement which is still used.

A 'yard' is slightly shorter than a 'metre'.

5-a-side team

You decide to start a new 5-a-side team but have no players at the moment.

You have £100 to spend. You need to buy only 5 of the players below.

Under their names are the amount each player costs to buy.

John	Raj	Jav	Nick	Paul	Wilma	Rob
£20	£10	£25	£20	£5	£15	£10

Write below the 5 players you want in your team, and the cost of each.

_____ = £ [ ]

_____ = £ [ ]

_____ = £ [ ]

_____ = £ [ ]

_____ = £ [ ]

How much did you spend in total? [ ]

How much of your £100 do you have left? [ ]

17

 Star Striker

United's star striker is called Don Deeds. Below are United's last 5 matches. The number of goals that Deeds scored in each match is in brackets.

United 3 - 2 Palace   (*Deeds* 2)

United 2 - 5 Vale   (*Deeds* 1)

United 4 - 1 Albion   (*Deeds* 4)

United 3 - 3 City   (*Deeds* 1)

United 0 - 1 Rovers   (*Deeds* 0)

> *Deeds* scored 2 goals in the first match

Against which team did Deeds score no goals?

Against which team did Deeds score all of United's goals?

Against which team did he score half of United's goals?

When we study numbers, such as goals or shots in a football match, these numbers are called **Statistics**

Jane and Paul played each other in a match. Here are their **Statistics**.

Jane		Paul
	Shots	
2	Shots	6
2	Goals	3
60	Passes	50

50% is half, 100% is all

Can you see above that Jane had 2 Shots and scored 2 Goals!

Did she score from 50% of her shots or 100% of her shots? ☐

Paul had 6 Shots and scored 3 goals.

Did he score from 50% of his shots or 100%? ☐

19

Football Shirts

Multiples of 10 include:
10, 20, 30 ,40, 50, 60, 70, 80…

"**Round up**" or "**Round down**" the numbers on the kits to the nearest multiple of 10. If the shirt number ends in '5', then round up, for example 15 rounds up to 20, not down to 10.

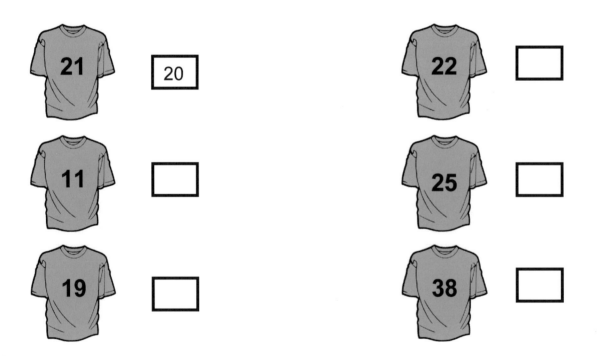

21 → 20

22 → ☐

11 → ☐

25 → ☐

19 → ☐

38 → ☐

Score a goal by working out which numbered square to shoot at. Calculate the answers to the puzzles below. Each answer matches a number in the goal. Cross each number off the picture as you find it, until only 1 square is left.

4 - 3 = ☐

5 + 2 = ☐

9 – 4 = ☐

2 x 2 = ☐

2 x 3 = ☐

Half of 4 = ☐

50% of 16 = ☐

Tick the box next to the shirt numbers which are multiples of **4**.

4 ☐

12 ☐

6 ☐

18 ☐

24 ☐

16 ☐

20 ☐

8 ☐

 Which shirt?

Your new manager is going to assign a shirt to you. Work out the puzzles below to find out which shirt number you are going to be given.

 Write an even number between 10 and 20 in the box

 Divide it by 2

 Take away 4 from your number

 Multiply your number by 2 to give your shirt number!

23

 League Tables

In football you need to learn about league tables

Villa **won** their first match of the season. The league table looks like this:-

Team	Played	Won	Drawn	Lost
Villa	1	1	0	0

Villa then played a second match. The league table now looks like this.

Team	Played	Won	Drawn	Lost
Villa	2	1	0	1

Look at the table and work out if Villa won, drew or lost their second match.

Villa **won** their third match of the season. Can you fill in the blanks?

Team	Played	Won	Drawn	Lost
Villa	3			

 League Calendar

Below is a calendar which has 2 days missing, one in blue, one in green.

May

Sunday 1	Monday 2	Tuesday 3	Wednesday 4	Thursday 5	Friday 6	Saturday 7
Sunday 8	Monday 9	Tuesday 10	Wednesday 11	Thursday 12	Friday 13	Saturday 14
Sunday 15	Monday 16	Tuesday 17		Thursday 19	Friday 20	Saturday 21
Sunday 22	Monday 23	Tuesday 24	Wednesday 25	Thursday 26	Friday 27	
Sunday 29	Monday 30	Tuesday 31				

The day that is highlighted in blue is the last league match of the season. Work out what **day** and **date** should be in the blue square, then write it in.

The day highlighted in green is the cup final! Work out what **day** and **date** should be in the green square and then write it in.

Answers

Page

1 (a) Spain (b) 9th (c) Italy (d) England (e) Netherlands (f) Argentina

2 (a) 1 (b) 7 (c) 5

3 Town = 6, City = 3, Rovers = 3, Albion = 0

4 12, 8, 16, 24, 4, 16, 12, 20, 8, 12, 24, 20, 4

5 (a) 2 (b) 4 (c) 0

6 (a) +1 (b) -1

7 (a) City 2-1 North End (b) City 1-4 United

8 (a) 3 (b) +1 (c) -2 (d) City

10 4, 8, 4, 12, 20, 24, 12, 20, 12, 24, 20, 8

11 (a) 1 (b) 1½ (c) 2½ (d) 3

13 7

14 10, 20, 10, 20, 30, 30, 40, 20

15 20, 10, 10, 20, 20, 30

16 (a) 12 (b) 8

18 (a) Rovers (b) Albion (c) Vale

19 (a) 100% (b) 50%

20 20, 10, 20, 20, 30, 40

21 (a) 1 (b) 7 (c) 5 (d) 4 (e) 6 (f) 2 (g) 8. So the square = 3

22 4, 24, 20, 12, 16, 8

24 (a) Villa lost their second match

25 Played 3, Won 2, Drawn 0, Lost 1

Other books by Adrian Lobley

The Football Maths Book

The Football Maths book is a fun way for children to improve both their maths skills and their soccer knowledge. They will learn about player positions, formations, cup competitions, penalty shoot-outs, own goals, hat-tricks and many more football terms. At the same time they will be counting, adding, subtracting, halving, doubling and learning odds and evens. The Football Maths Book is a Key Stage 1 maths teaching aid for 4-7 year old football fans.

The Football Maths Book The Rematch!

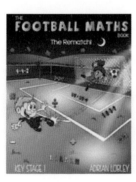

"The Football Maths book - The Rematch!" is the sequel to "The Football Maths Book" and is a fun way for children to improve both their maths skills and their soccer knowledge. They will learn about league tables, football stats, goal difference, diamond formations, substitutes, red and yellow cards and many more football terms. At the same time they will be learning about multiples, fractions, greater/less than, pentagons, hexagons, clock times and many more. "The Football Maths book - The Rematch!" is a Key Stage 1 maths teaching aid for 5-8 year old football fans.

A Learn to Read Book: The Football Match

The Football Match is a book to help young children read their first words. It is targeted at young football fans who know the alphabet but cannot yet translate the combination of letters into words. It also assists with writing, numeracy, selecting the correct page and learning about football.

The book is an adventure where children choose on each page what they want to happen in the football match. When they score a goal they write the new score in. Developed to link school and home learning, 'The Match' series builds children's confidence and also introduces them to phonics.

Printed in Great Britain
by Amazon